How to teach this unit

How often have you remonstrated with your pupils for gazing out of the window during a lesson? It's not easy to compete with the sight of winter's first snowfall, or even a noisy mower, especially when the alternative is the nitty gritty of spelling rules in a stuffy classroom!

Well, this unit should appeal to you and your pupils, as it will actively encourage some window-gazing! This window-gazing, though, is active and purposeful: carefully thought-out activities which develop important geographical skills and understanding (see **concept map**). Care has been taken to select a wide range of different photographic images to encompass views showing:

- different perspectives: from ground level, low oblique, and (to scan a wider area) higher oblique and aerial shots

- different locations: familiar urban and rural settings from around the UK, and more distant places – European and worldwide

- different degrees of environmental quality.

Time allocation

The teaching time available for geography can vary enormously. SuperSchemes units have been written with three possibilities in mind:

- a **short–medium** unit (5–10 hours)

- a **long** unit (10–15 hours)

- a **continuous** unit (15–30 minutes per week).

The medium term plans allow you to choose an appropriate length for your particular class. Some of the longer medium term plans offer enough material for you to continue with the topic later in the year.

What do I need to know?

It's interesting that when prospective buyers are appraising a property for the first time, they often look straight past the room and gaze out of the window to assess the view. The view from our windows is very important to us, and can evoke different responses – sometimes it is an immediate gut reaction of affinity or dislike, and at other times it may just evoke a feeling that is hard to define or explain. What is also interesting is that some people's ideal view – perhaps an open expanse of moorland, or

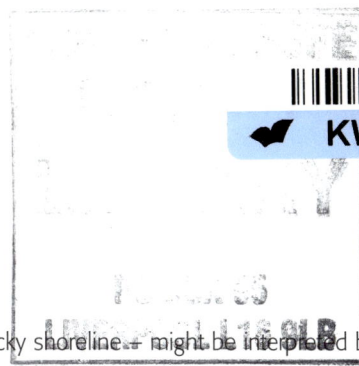

a rocky shoreline – might be interpreted by others as a bleak, unfriendly and even frighteningly quiet environment in which to live. Conversely, a view of a lively and crowded city centre which appeals to some people might be someone else's vision of hell!

It's important to remember this when presenting children with different images of views – our pre-conceived ideas of what the children are likely to view positively may be wrong. Depending on their personalities and experiences, children's perceptions may vary enormously. One of the appeals of this unit is that it gives children the opportunity to develop and express their thoughts and feelings about both familiar and unknown environments, in the safety of a non-judgemental setting. Importance should be attached, not to the responses of the children to a particular view, but to their growing ability to analyse their responses through your use of carefully planned questioning (see Using visual images on the CD).

Depending on their age and ability, different children will demonstrate different levels of sophistication. In their simplest form, these levels can be described as the ability to **observe** → **describe** → **explain** → **analyse** the view in question.

Key ideas

In addition to providing the context for geographical activities, this unit can help to develop pupils' specific thinking skills via the use of Edward De Bono's model of coloured 'thinking hats'. Activities that favour a particular style have been incorporated into the suggested pupil tasks in the **medium term plan**.

De Bono identified six distinct types of thinking, all equally valid and important, and represented them by coloured hats:

- A **white hat** represents information and facts.

- A **red hat** represents intuition, feelings and emotion.

- A **green hat** represents creativity, thinking of suggestions and alternatives.

- A **yellow hat** represents thinking of positives, advantages and benefits.

- A **black hat** represents judgement, caution and logic.

- A **blue hat** represents thinking itself, i.e. thinking about thinking – 'what have we learned/found out/decided?'

If your pupils are unfamiliar with this approach to their work, it will be helpful to prepare the ground. A visual aid (wall display, hanging mobile, actual coloured hats) will help the children to begin to associate each colour with a particular style of thought. The pupils could also assign each colour an appropriate image, e.g. grass growing for green thinking. Then you can introduce a problem-solving situation which uses each skill/coloured hat in turn. Alternatively, you could use the story *The Magic Hats* by Lorna Santin; for this, and for more information on de Bono's thinking hats, follow the **Resources** link from the CD.

Field sketching

This is a key skill that pupils will develop and practise throughout this unit. Year 4 children may well have a huge range of abilities in this area, depending on previous experiences and general sophistication of motor and observational skills. For some pupils, having to cope with a clipboard in a breeze can be extremely taxing.

It is important to remind children that a field sketch needs to concentrate on the general composition of the view and relative position of items within it – they are not being asked to produce a work of art. Many children become obsessed with the intricate details of individual objects and fail to notice the bigger picture.

Drawing the view from your classroom window by sitting *indoors* initially would be a good way to introduce this skill, as the very frame of the window itself will give a clear boundary to their sketches. Once confident in this, the children can move outdoors. If some children find it very difficult to get started, or keep rubbing bits out, give them a few key boundary lines to get them going – e.g. the edge of a paved/grassy area.

Suggesting that the children leave a ruler's-width blank border all the way round the page is a useful way of enabling them to add annotations and labelling. Some children may also need reminding not to let their labelling lines cross over, as this leads to confusion. Encourage children to include the location and compass direction of their view. Some more able pupils may choose to identify features within their view using a key.

If appropriate, model a way of approaching a field sketch yourself, and talk through what it is that you are doing and why, e.g. 'I can see that there is an overflowing rubbish bin just in front of the wall here, so I'm going to draw a cylindrical shape with some rough squiggles around it and then label to the nearest side what it is so that I remember later on.'

On the CD there is an example of a field sketch done by Year 5 pupils.

Where do I start?

To prepare for teaching this unit, it would be helpful to check the **medium term plan** so that you know which lessons take place outside the classroom and can plan in any necessary additional support.

It will also be useful to collect together the following resources:

- a set of photos particular to your school setting, showing views from different windows, ready for use by pairs of children

- downloadable activity sheets and photos from the CD

- school plans and aerial photographs

- holiday snaps of views – ask colleagues and parents for these, stressing the point that 'ugly' ones, such as a building site below a balcony view, would be especially welcome!

Introductory activities

Examine the use of the word 'view' with your children – it can have two different meanings, both of which will be considered in this unit:

- what you can see in front of you

- your opinion of something.

Show the children a photograph of a view – what everyone sees is the same, but their interpretation of it will differ. Remind them that everyone is entitled to their own view, and that no one is 'right' or 'wrong', although there may be a consensus of opinion.

Introduce the children to the ideas of perspective and scale – allow them to look out of the classroom window. What happens if they crouch down? Would they see the same view from the ceiling? What sort of view would a bird get? What about a high-flying jet, or even an orbiting satellite?

Perhaps set a weekend homework task to draw a view from a window at home – this could be the start of an evolving, interactive classroom display.

Ask the children to begin to collect colour photographs of any type of view for their own visual resource bank.

Discuss what type of structures have windows (and therefore a view from them): how might their views differ? For example, a bungalow window, a window upstairs in a house, or on the 10th floor in a block of flats. What about windows in a car, an aeroplane or a ship? Would castles, cottages and mud huts have the same view? Show them some photos of windows from the CD to get their imagination going.

Use the pairs of photos in the photo-bank 'Match the view to the window' on the CD, to see if the children can correctly identify the location that each view was taken from.

Concluding activity

As a final activity for this SuperSchemes unit, a good way to broaden your pupils' perspectives is to ask them to find out about one of the many local, national or international organisations or charities whose job it is to look after our view by preserving and conserving the country's buildings, countryside and different environments. The National Trust (www.nationaltrust.org.uk), English Heritage (www.english-heritage.org.uk), the Environment Agency (www.environment-agency.gov.uk) and EnCams, which runs the Keep Britain Tidy campaign (www.encams.org/home/).

This might make a good homework web research activity:

- Can the children find out what they do and how?

- Is there an organisation based near your school that looks after your local area – what project has it been working on recently?

- Do they need any volunteers to help out? Could the children get involved?

A local newspaper is a good source of information and contact details.

Medium term plan:
Improving the view from our window: Do you see what I see?

Learning outcomes	Key questions	Pupil activities	Resources/ Key vocabulary	Assessment opportunities
To recognise that a 3D object will look different when viewed from different angles	What do objects look like from different positions?	Ask the children to draw a collection of 3D objects from different angles – from above, below, eye-level, obliquely – using different colours for different faces. _Classroom-based, individual practical activity_	Collection of 3D objects **Activity sheet 1** _view, angle, position, eye-level, bird's-eye, oblique, aerial_	Can the children complete **activity sheet 1**?
To translate 3D shapes into plan format	What do these objects look like from above?	Invite the children to create a small 3D 'scene' or collection of shapes (using any construction equipment) and make a 'bird's-eye view' drawing of it. Differentiation: BA: take oblique and vertical photos to assist 3-D drawing. AA: use squared paper to begin to understand scale (smaller squares for drawing than base for 'scene') Give the children OS maps to look at – how are features represented as symbols in the key? _Classroom-based, individual **and** group practical activity_	Different sizes of squared paper Construction equipment 3D shapes Digital camera OS maps _plan, map, symbol, key_	Can the children translate shapes into different perspectives?
To locate and match familiar scenes with their viewing point	Where is this view? Where was it taken from?	Show views from inside the school looking out – can the children identify the view? Devise a school trail: using thumbnail versions of each photo, small groups of children find where each photo was taken from. Get groups of children to start from different photos, so everyone has to think for themselves rather than just follow others. _Out of classroom (but inside school building) small group or pair activity_	Digital camera Series of photos shot from school windows _location_	Can the children find the correct locations? Which were easiest/ hardest and why? What camera angles were used?
To gather information (**white hat thinking** – see **key ideas** in **How to teach this unit**) To express an opinion about a site To recognise that different people have different values and points of view (**red hat thinking**)	Do I like my view?	Use the view from the classroom window to carry out the environmental quality survey (activity sheet 2). Once children are confident with the process, groups of six can select one of the photo trail locations for an in-depth study. Ensure that all pupils realise that there are no right or wrong answers. TA to check that BA children understand how to complete the survey. _Whole class, then group fieldwork in school grounds_	**Activity sheet 2** **Resource sheet 2** from CD for recording group responses **Lesson plan** _environmental quality, survey, opinion, judgement, rating, score_	Can children analyse their group's data to determine which location was most/least liked? What were the main features of the site that determined this?
To recognise that people can change their environment in a positive way (**green hat thinking**)	Does my view need improving?	Analyse and present the survey results. How do the different groups' results compare? Revisit the 'worst' site, draw an annotated field sketch and consider how this view might be enhanced. Would this involve a physical change to the environment, (e.g. removing or adding a structure) or to the way we use it (e.g. changing a football pitch into a quiet area with seating)? Consider how you might follow up the children's findings and suggestions. _Group based discussion leading to whole class presentation/ feedback_	Survey results Data-handling package _site, field sketch, annotate, habitat, manage, improve, change, biodiversity_	Can the children make appropriate suggestions for possible improvements to the 'worst' location?
To (a) recognise that our environment is constantly changing, and (b) understand the different timescales involved in physical processes and human impacts (**yellow and black hat thinking**)	How has this view changed? Is change always an improvement? How might local people view the changes?	In pairs, ask the children to study the pictures on **resource sheet 3**, looking for changes over 50 years. Are all of them improvements? Would everyone agree? Give each pair of children a prepared window frame, and ask them to draw what the same view might look like in the future.	**Activity sheet 3** **Resource sheet 3** A4 sheets with a simple window frame _change, improvement, opinion, human, physical, impact_	Can the children differentiate between changes to the environment caused by humans, e.g. road building, and physical changes e.g. growth of vegetation?

Medium term plan:
Improving the view from our window: Do you see what I see?

Learning outcomes	Key questions	Pupil activities	Resources/ Key vocabulary	Assessment opportunities
To use photographs to describe what a place is like (**white and red hat thinking**)	What does this view tell me?	Ask the children to be 'geographical detectives', looking for clues by 'reading' photographs. Children could generate these questions themselves: ■ What can you see? ■ Is anything happening? ■ What might people be saying/thinking? ■ What might the same view look like at different times of the day/year? ■ How does this picture make you feel? ■ What do all the photos have in common? ■ Where might the photographer be? How do you know? Ask the children to group the photos using their own criteria, such as: like/don't like, this country/abroad, rural/urban. Compare and discuss differences. *Small group, photo-based classroom activity*	Photos of the UK and the world from the CD	How accurately can children 'read' a photo? Can they draw out-of-shot extensions to the right/left?
To analyse secondary evidence and draw geographical conclusions (**white and green hat thinking**)	What clues to a view does the style of window give?	On the class board draw three empty window frames, e.g. Georgian bow, Victorian sash, modern picture window. As a class, discuss what sort of view you might see through them – urban or rural, UK or abroad, present-day or historical? Invite the children to record their own interpretations inside the frames using **resource sheets 4a–h**. Encourage BA children to select frames they are familiar with. Ask AA children to select a frame and draw the building it belongs to, including someone in the window looking out. *Class discussion and teacher modelling, leading to individual recording*	Pictures of different types of window from **resource sheets 4a–h** *modern, urban, rural, local, distant, foreign*	Are there any obvious misconceptions or visual/geographical anachronisms?
To interpret photographic images from other people's points of view (**red, yellow and black hat thinking**)	Who saw this view? Where were they? What were they thinking/feeling? Why did they take the photograph?	In pairs, ask children to play a selection of characters: senior citizen, newspaper reporter, tourist, themselves. Ask each pair to have a conversation that results in them taking each of the photos, recoding their conversations on **activity sheet 5**. In a plenary, ask them to act out their conversations, and compare variations. Pair work – drama activity	Photos of the UK and the world from the CD **Activity sheet 5** *Key vocabulary will depend on photos and individuals selected*	Could the children empathise with different people's points of view?
To summarise what has been learned, using appropriate vocabulary (**blue hat thinking**)	How does perspective (visual or attitudinal) change a view? Does everyone 'see' things in the same way? What have I learned?	Remind children of all the work that they have done, including those activities with no paper recording. What do they think they have learned? What did they enjoy the most? What skills do they think they have improved and will be able to use again? Complete **activity sheet 6**.	**Activity sheet 6** *thinking skills, progress, confident, enjoyed, assess, improve, future, target*	Final self-evaluation assessment using **activity sheet 6**

Cross-curricular links:
PHSE and citizenship – the chance to express opinions about their school environment, to become involved in a project to improve it, and to extend their empathy skills; **maths** – spatial awareness, use of 3D shapes; **art** – looking at artists who have been inspired by the view from a window, e.g. Magritte, Matisse, Picasso, Dürer and Hockney; **English/drama** – role-play, providing contexts for extending speaking and listening skills; **history** – ideas of continuity and change in local environments; **ICT** – use of data-handling packages to present survey data.

Concept map

(skills-based)

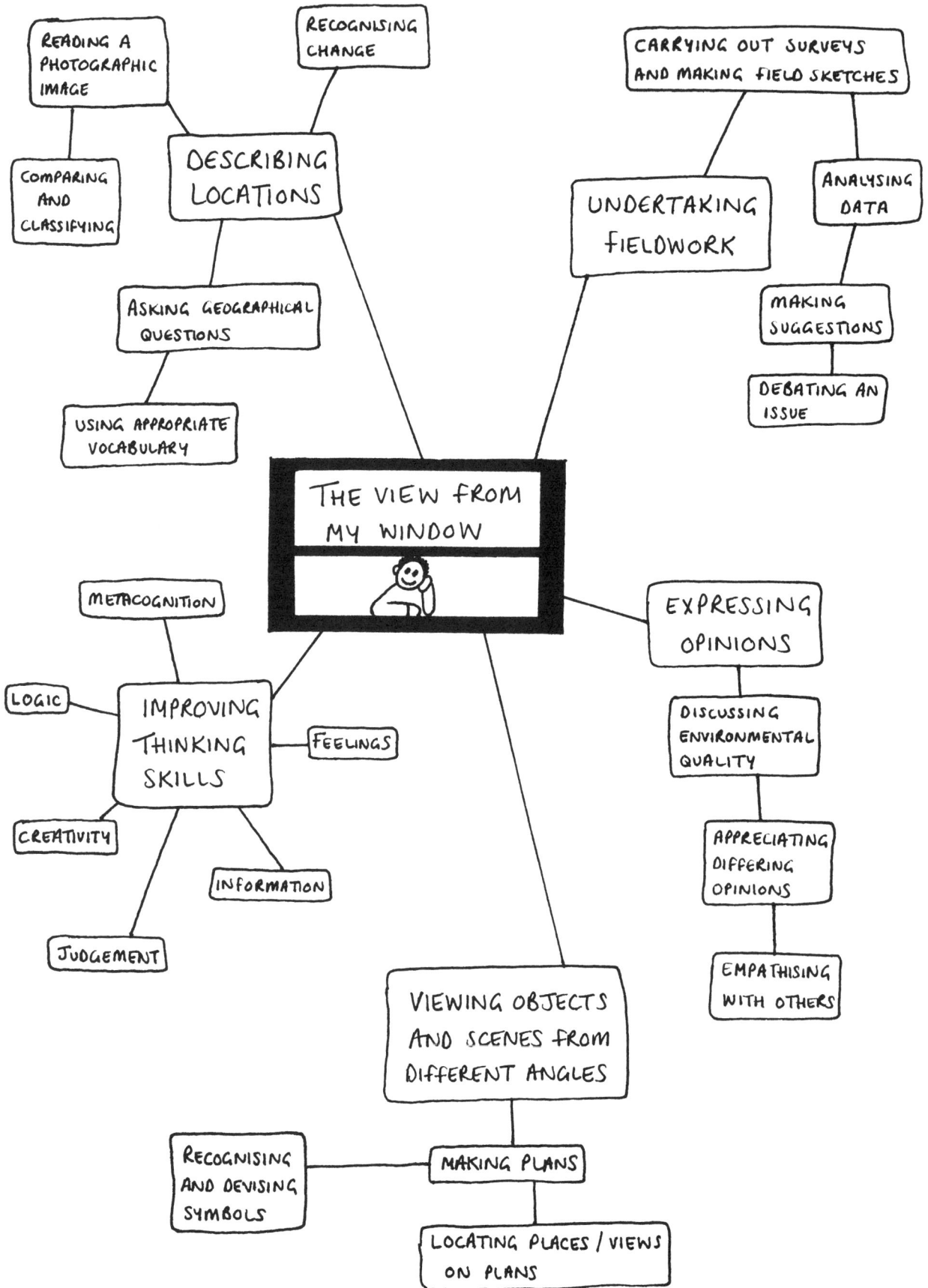

READING A PHOTOGRAPHIC IMAGE

RECOGNISING CHANGE

COMPARING AND CLASSIFYING

DESCRIBING LOCATIONS

CARRYING OUT SURVEYS AND MAKING FIELD SKETCHES

UNDERTAKING FIELDWORK

ANALYSING DATA

ASKING GEOGRAPHICAL QUESTIONS

MAKING SUGGESTIONS

USING APPROPRIATE VOCABULARY

DEBATING AN ISSUE

THE VIEW FROM MY WINDOW

METACOGNITION

LOGIC

IMPROVING THINKING SKILLS

FEELINGS

EXPRESSING OPINIONS

DISCUSSING ENVIRONMENTAL QUALITY

CREATIVITY

INFORMATION

APPRECIATING DIFFERING OPINIONS

JUDGEMENT

EMPATHISING WITH OTHERS

VIEWING OBJECTS AND SCENES FROM DIFFERENT ANGLES

RECOGNISING AND DEVISING SYMBOLS

MAKING PLANS

LOCATING PLACES / VIEWS ON PLANS

Lesson plan: Do I want to change my view?

Subject: Geography (year 4)

Time/Duration: Two hours +

Learning outcomes

In this lesson children will be developing:

Red hat thinking (feelings)

- to express an opinion about a site
- to recognise that different people have different points of view

White hat thinking (information)

- to gather information about a location by creating, using and analysing a survey

Black hat thinking (negatives)

- to filter out ideas that are not practical, sensible or achievable

Green hat thinking (ideas)

- to recognise that people can change their environment in a positive way.

Background to the current lesson

This is the fourth lesson from the medium term plan, and follows on from the photo trail around the school, where views from different windows were recognised and located on a plan.

Lesson details

Introduction

As a class, ask the children to close their eyes, 'put on' their red thinking hats, and picture a real or imaginary lovely place. Tell them to wander around the place in their minds, looking up, down and around. What can they smell and hear? How does the place make them feel? When they open their eyes, ask them to think of adjectives to describe their place and write them on a whiteboard.

Ask them to repeat the exercise, but for an unpleasant place. When they think of adjectives, suggest they look for pairs of rough opposites, e.g. dull/colourful, frightening/friendly. Show them how these opposites could be placed on a numbered scale, with 1 most negative and 5 most positive.

Choose three different and contrasting scales to test in the classroom. Ask the children to rate each environmental quality. Remind them that there are no right or wrong answers – likes and dislikes are personal. Use a tally to collate the number of hands raised for a score of 1, 2, 3, etc. Which rating got the highest number of votes for each of the categories?

Main activity

Explain to the children that they are going to follow a similar set of steps for collecting evidence about the environmental quality of different areas of the school (white thinking hats). Divide them into groups of six, either in mixed-ability groups or, if a TA is available, in ability groups. Give each group a different photograph from the previous lesson's trail (obviously omitting or being cautious about any potentially dangerous sites, such as roads or car parks). This time they are to go and stand in the view itself (i.e. outside) to conduct the survey. Remind them to look up, down and all around, and use all their senses. Ensure that all the children understand how to fill in the recording sheet (**activity sheet 2**), and ask them to complete the location details before they set off. Remind them not to be influenced by what other children think.

When each group returns, ask them to collate their results, give a rating for each quality using the tally chart (**resource sheet 2**), and make a statement describing the environmental quality of their area.

Plenary

Each group gives feedback to the rest of the class. They could show their average scores on bar charts so that the 'best' and 'worst' locations can be identified. Visit the 'worst' site and make annotated field sketches on A3 paper, labelling features and adding comments (white thinking hats again). Back in the classroom, change to green hat thinking and discuss how the area could be improved. You may need to sound a note of caution here (black hat thinking) as not all the ideas may be practical or financially possible!

Some suggestions could be opened up to a wider audience – governors, school council or eco-school committee. An action plan could be drawn up, grouping improvements according to their rough cost. It's especially worthwhile if children can see that they are taken seriously and can make a difference to their environment. Ask them to research organisations which protect our natural and built environment (see **Resources** section on CD).

Differentiation

Expect AA children to offer more detailed field sketches and explanations.

Resources

Each group should have a set of photos from the school trail.

Cross-curricular links

- **PSHE and citizenship** – expressing an opinion about the school environment and being involved in a project to improve it.

Using the activities

Activity 1: Who can see which view?

Children will have met these shapes in their maths lessons, so this activity will build on existing knowledge, encouraging them to look at their environment in different ways and to recognise that views can change depending on your position when you look at them. It also introduces children to the idea of vertical and oblique aerial photographs; when they first encounter such photographs, it will be useful to remind them of this exercise. It will also encourage them to be more creative in drawing their own views in later activities.

Activity 2: Do I like my view?

Often children are not naturally observant: it is a skill they have to learn. You will need to encourage them to think about the quality of an environment, rather than uncritically accepting what is there. This activity will also help them to understand that individual views about a place will be different, and that it is quite acceptable to think differently from other people – the point is to be able to justify and give evidence for their views. Supported by clear reasoning, different views can be seen to be equally acceptable. This will be a good discussion point, as children often dislike being different and like to have a neat answer – rather in the way their maths problems might be resolved.

Activity 3: How has this view changed?

This activity provides an excellent opportunity for looking at the process and results of change over time. Display the pictures on Resource sheet 3 on the IWB, or print off enough copies for groups of children. Give each child a copy of Activity sheet 3. The two 'then and now' pictures are 50 years apart, but children will also be aware of changes occurring in their local area, and these will make a good preliminary discussion before the activity.

Some children are more comfortable with the known, familiar present than the idea of the past, so it is important to talk about the nature of change, how it happens and whether or not the changes always result in improvements to the environment. Again, individual perceptions will differ; children will be interested to discuss the idea that different age groups might hold different opinions about change.

Trying to imagine how a place might change in the future is a very good exercise. Ask the children to look at the changes that have taken place in the 'now' picture and think how that might change in the future. What will be the big issues? Transport, housing, leisure activities?

Activity 4: Draw my view

This presents children with an exciting task, requiring them to look at window frames of different types and from different periods and to predict the type of view they might expect to see through them. To get them started, you might find it useful to look at pictures showing a variety of views and locations, and discuss the contexts for these, to help the children think more widely about the views they might see from their own windows.

Remind them of the types of view they might see. For example, in a car wing mirror the view will be of what's behind the car, while an aeroplane window might show a view of the land below. Encourage them to think imaginatively about global locations as well as local views.

Activity 5: Why are they taking this picture?

This is an opportunity for some role play centred on why different people might take photographs. Some preliminary discussion will be needed to help children enter into the role play, and they will need to understand that the views of the people they play may be different from their own. Some reasons to start them off might include:

- to make a record – store a memory
- to share with others
- to record something such as piles of litter or something broken as evidence for a campaign
- to provide illustrations for a newspaper article or a magazine.

Activity 6: What do I know now?

This is a chance to enable children to reflect on, and record, the work they have done and the skills they have mastered. Discuss with them the need for an honest assessment of their achievements, and help them to understand it as a foundation for future work rather than a judgement on what they have done.